# Pythagoras'
## (2500 year old)
# Secret Code

This triangular arrangement of 10 dots is Pythagoras' secret code. Within it, he saw (and heard) the "harmony of the spheres."

He made his followers in the *Order of the Pythagoreans* swear their oaths of secrecy on this shape.

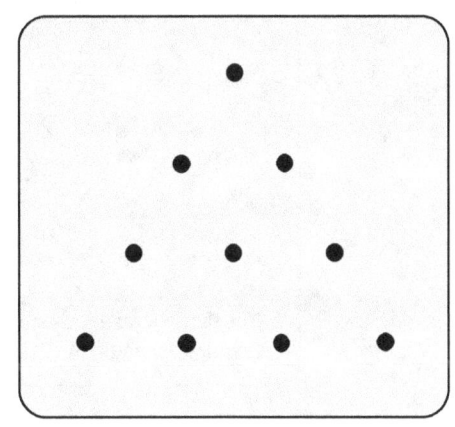

PYTHAGORAS

Pythagoras saw a lot more in this set of dots than simply a way to arrange bowling pins. Concealed in this simple configuration is a series of harmonies that can be found in **Arithmetic, Geometry**, and **Music** (and Pythagoras included **Astronomy** as well.)

But before revealing how this shape explains an amazing connection between **number, shape**, and **sound**, (and **planets**) let's first learn a little about this genius was who was born in 570 BC.

## The first math whiz of the Western world.

In his youth, Pythagoras traveled from his native Greek island of Samos to Athens, Egypt, Crete, and even to Babylon in Mesopotamia, visiting the world's wisest mathematicians.

He finally settled in the Greek colony of Croton, on the heel of Italy, where he set up a school to share what he had learned.

(Incidentally, his name sounds much cooler in Italian: *Pitagora*, pronounced "PEET- a- gorrra.")

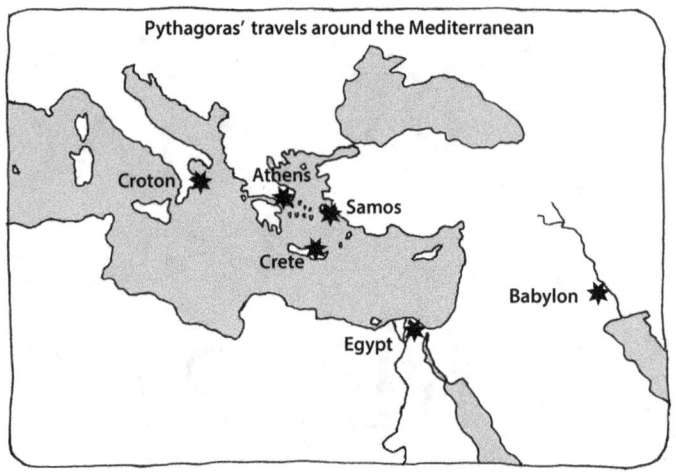

Pythagoras' travels around the Mediterranean

## The shape ruled heaven and earth.

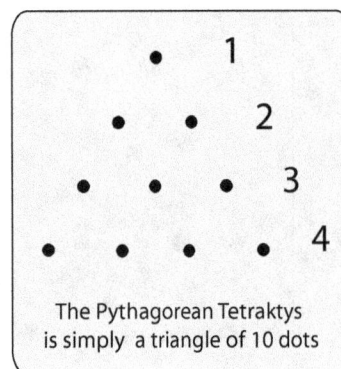

The Pythagorean Tetraktys is simply a triangle of 10 dots

Pythagoras called his sacred shape the *tetractys*, which basically means "a group of four things." The four horizontal rows represent the digits one, two, three, and four.

Pythagoras liked other "groups of four things":
Elements (fire, air, earth, and water)
Seasons (spring, summer, fall, winter)
Magnitudes (point, line, surface, and solid)
Ages of a Person (infancy, youth, adulthood, old age)

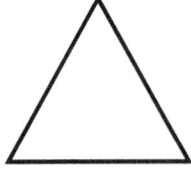

What's so special about one, two, three, and four? Well, as stacked dots, they make a nice equilateral triangle. Numerically they're important because they add up to 10, which is the base of our numbering system.

**10**

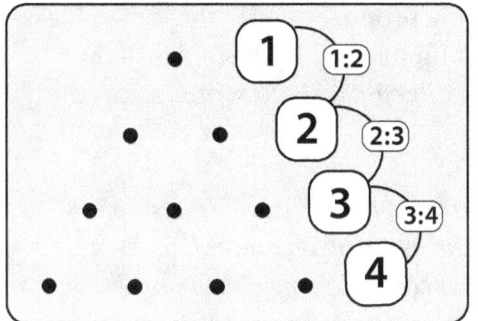

What Pythagoras really wanted us to see was not the numbers themselves, but the interrelationships **between** the numbers.

He wanted us to see the ratios **1:2, 2:3, and 3:4.**

Okay, so what's the big deal with a couple of ratios?

Well, realize that the Pythagoreans thought number was the "first thing in the whole of nature" and that "all things seem to be modeled on numbers" (according to Aristotle, writing around 350 BC).

So the interrelationships among these first four numbers was one of the most basic laws of nature!

> **By him who handed to our generation the tetractys, source of the roots of ever-flowing nature."**
>
> The Oath of the Pythagoreans (according to Iamblichus, around 300 AD)

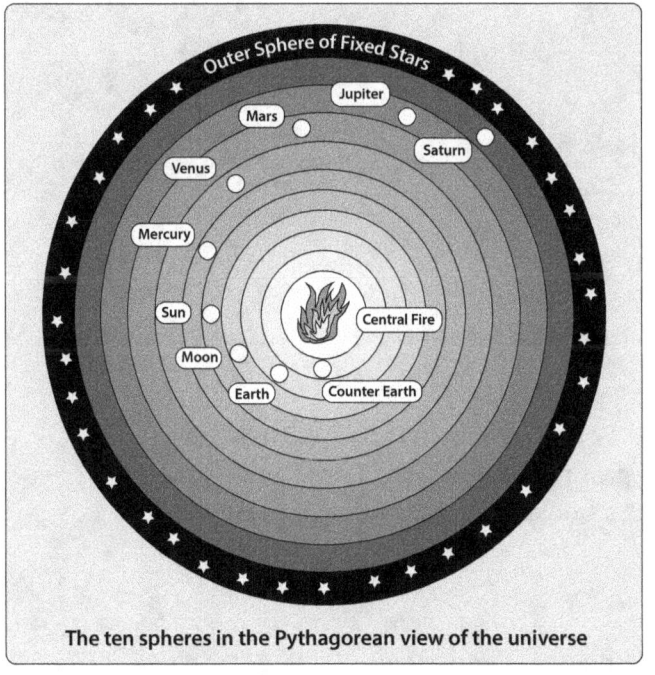

The ten spheres in the Pythagorean view of the universe

Pythagoras thought the cosmos consisted of 10 spheres swirling around a blazing fire.

The inner spheres made a certain sound whizzing through the air. The middle spheres moved faster and made a higher pitched sound. And the outer spheres made a really high pitched sound.

The distance the spheres were from the fire and thus, the speed of their whizzing through space, created the "harmony of the spheres."

Pythagoras and his followers set out to refine this model of the "mathematical Universe" by studying the mathematics of music.

This all might seem like some silly old man making up science fiction number games. But numerous heavy hitters throughout history have seen the magic in the tetractys, and realized Pythagoras was on to something important.

### *Plato spreads Pythagoraean wisdom*

Around 325 BC, Plato was a big proponent of Pythagorean ideas on number and music. Plato even set up his own Pythagorean-like school in Athens. In his famous book, *Republic*, Plato writes about the harmony of the spheres:

*"and up above on each of the rims of the circles a Siren stood, borne around in its revolution and uttering one sound, one note, and from all there was the concord of a single harmony."*

(A Siren is a mythological creature who sings.)

### *Virtuous Vitruvius*

In 25 BC, the Roman architect Vitruvius explained that in large amphitheaters, hidden beneath the seats of the audience, were huge metal bells shapes that would echo and amplify the sounds of 1:2, 2:3, and 3:4 coming from the stage. *(More on these sounds and Vitruvius later.)*

The "sounding vessels" were placed in small chambers scattered throughout the theater to improve acoustics.

Greek "sounding vessel", as described by Vitruvius

### *1:2, 2:3, and 3:4 among the Neoplatonists*

Around 125 AD, Nichomachus, a "Neo-Platonist" philosopher (a group that revived Platonic and Pythagorean thinking) wrote a book called *Introduction to Arithmetic.*

In 525 AD, the Roman author named Boethius translated Nichomachus' Greek *Introduction to Arithmetic* into Latin.

*Introduction to Arithmetic* was the most widely used mathematics text up until the Renaissance. That's a run of over 1300 years! (Modern books are lucky if they last a month on the New York Times bestseller list.)

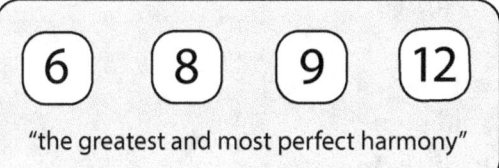

"the greatest and most perfect harmony"

On the very last page of *Introduction to Arithmetic*, both Nichomachus and Boethius reveal that the **"greatest and most perfect harmony"** involves the numbers 6, 8, 9, and 12.

What's the big deal with these numbers?
How come 7, 10, and 11 are left out?
The clue goes back to the tetractys.

Again it's the relationship between the numbers that's important.

The ratio of 6:12 is equal to 1:2.
The ratio of 6:9 is 2:3.
And the ratio of 8:12 is also 2:3.

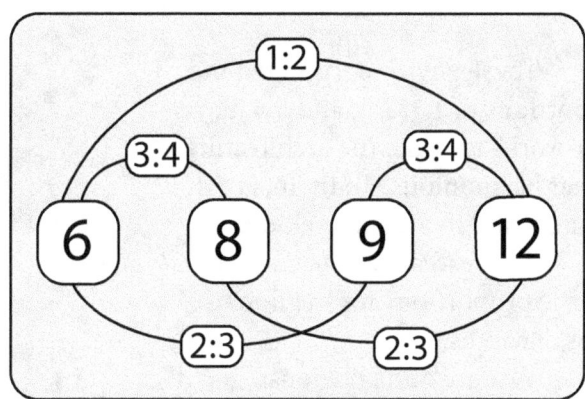

But wait there's more.
The ratio of 6:8 is 3:4.
In the ratio of 9:12 is also 3:4.

So, hidden among these numbers are Pythagoras' killer ratios, 1:2, 2:3, and 3:4.

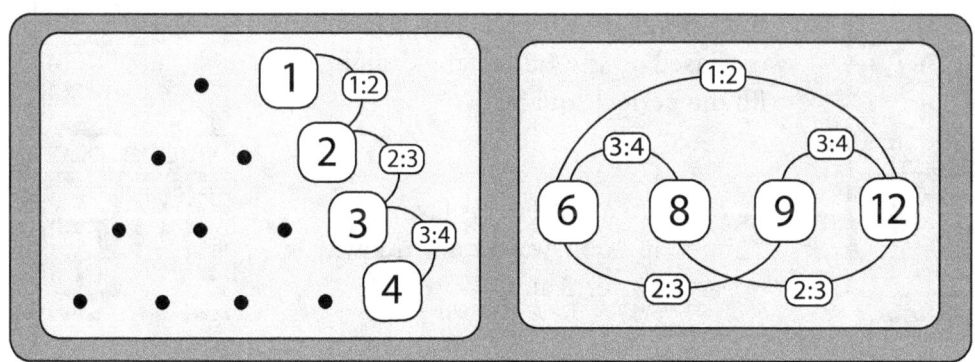

## 1:2, 2:3, and 3:4 in the Renaissance

Flash forward to the year 1414. An Italian scholar with the poetic name Poggio Bracciolini was rummaging around an old monastery in France when he found a manuscript copy of Vitruvius' *Ten Books on Architecture*, written around 25 BC.

He brought to light this important text which had been lost for 1400 years. It soon became the groundwork for a revival of classical architecture in Italy, Europe, and eventually across the Western world.

VITRUVIUS

## Alberti and Palladio

LEON BATTISTA ALBERTI

In 1452, Poggio's fellow countryman, Leon Battista Alberti wrote his own *Ten Books on Architecture*, based on Vitruvius' work. Alberti influenced a whole slew of famous architects like Cesare Cesariano, Sebastiano Serlio, and Andrea Palladio.

These guys incorporated the proportions of 1:2, 2:3, and 3:4 in their works to make the architecture appear harmonious. To them, architecture could be frozen melody.

Andrea Palladio had seven favorite proportions for the floor plans of rooms.

Two are quite obvious:

the circle of the square.

Then there are those special ratios 1:2, 2:3, and 3:4.

The remaining two are the ratios 1: √2 and 3:5.

ANDREA PALLADIO

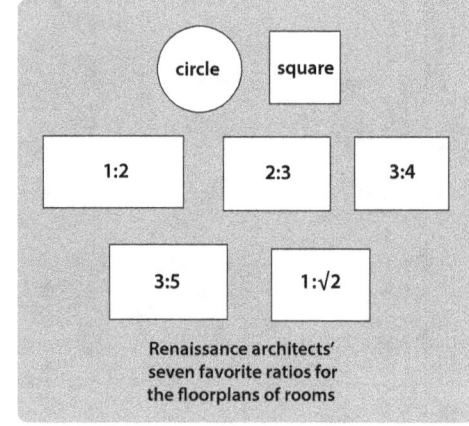

Renaissance architects' seven favorite ratios for the floorplans of rooms

Alberti's "1:2 plus a square" facade of Santa Maria Novella in Florence

Some of Alberti's church facades were based on the 1:2 rectangle, along with the perfect square.

In addition, his favorite ratio for a doorway or a window was a 2:3 vertical.

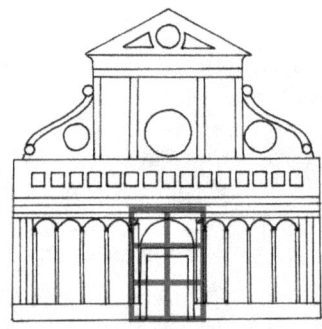

2:3 doorway of Santa Maria Novella, Florence

# *Raphael tucks the "greatest and most perfect harmony" into his famous 25-foot-wide fresco*

RAPHAEL

Around 1510, the Pope commissioned Raphael to paint a huge mural on the wall of the Vatican library. Raphael's *School of Athens*, depicts of the greatest stars of Greek antiquity.

Plato and Aristotle are in the middle. Various groups are clustered around Socrates (philosophers), Strabo and Ptolemy (astronomers), Euclid, (geometers), and Epicurius (pleasure-seeking scientists).

"THE SCHOOL OF ATHENS" by Raphael, 1509

Socrates lecturing · Plato holding *Timaeus* · Aristotle holding *Ethics*

Epicurius holding book · Pythagoras writing in book · Euclid holding geometers' compass · Strabo (or Zoroaster) holding celestial globe · Ptolemy holding terrestrial globe

Averroes (Arab translator of Greek works, ca.1175 AD) · Hypatia of Alexandria (famous woman mathmetician, ca. 400 AD)

Boethius holding his *Introduction to Arithmetic* with 3 book clasps visible on the back cover (ca. 525 AD)

Nicomachus peering (ca. 125 AD)

Pythagoras writing (ca. 500 BC) · diagram of "the greatest and most perfect harmony"

In the lower left is a group of mathematicians clustered around the burly Pythagoras.

Peering around Pythagoras' right shoulder is Nichomachus. Standing to the right of Pythagoras is Boethius. They both appear to be jotting down their versions of *Introduction to Arithmetic*.

Behind Pythagoras is the Arab mathematician Averroes (ca. 1150 AD) and the great woman mathematician Hypatia of Alexandria (ca. 400 AD).

close up view of
"the greatest and
most perfect harmony"

If you look closely, at Pythagoras' knees is a youth holding a tablet, which has the numbers 6, 8, 9, and 12, along with brackets and the words *Diapason, Diapente, and Diatesseron.*

These are simply the Greek terms for the ratios 1:2, 2:3, and 3:4.

Below the chart is the Pythagorean tetractys. And below that is in X, the Roman numeral for ten.

(At the top, the relationship between 8 and 9 is "one single note," which is called Epogdoôn, or "Tone.")

In short, Raphael is paraphrasing Nicomachus and Boethius, who got their wisdom from Pythagoras.

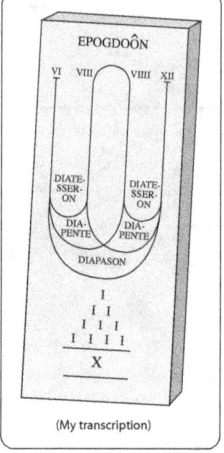

(My transcription)

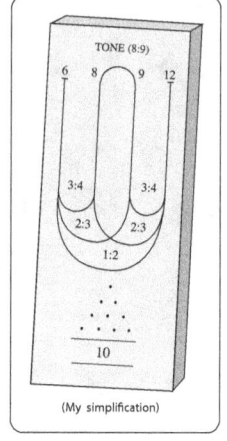

(My simplification)

## Franchinus Gaffurius teaches the musical harmonies

One of the most famous musicians in the Italian Renaissance was Franchinus Gaffurius. He counted Leonardo da Vinci as one of his good friends.

His book, *On the Harmony of Musical Instruments,* includes this wooodcut of Gaffurius himself, teaching musical theory to twelve students.

FRANCHINIUS
GAFFURIUS

Gaffurius is explaining, "Harmonia est discordia concors," which means, "Harmony is concord created from discord." The lengths of the flutes (Music) in the upper left are 3, 4, and 6. The lengths of the lines (Geometry) on the upper right are also 3, 4, and 6.

The interrelationships between 3, 4, and 6 succinctly encapsulate Pythagoras' ratios 1:2, 2:3, and 3:4. (As 3, 4, and 6 are simply Nicomachus' 6, 8, and 12, cut in half.)

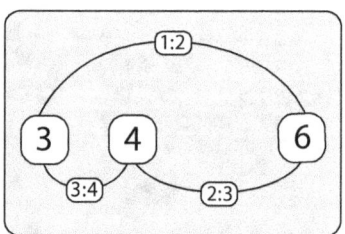

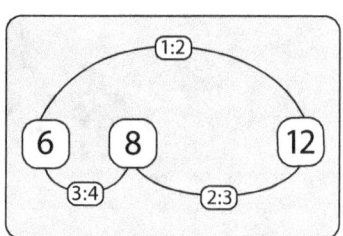

JOHN
DEE

*John Dee, Queen Elizabeth's philosopher, saw the magic in the ratios of Pythagoras' tetractys*

In the mid 1500's, the English mathematician, navigator and cartographer John Dee was also fascinated by the ratios 1:2, 2:3, and 3:4.

Dee was the first mathematician to use the colon in association with ratios. His first printed usage actually involves the 2:3 ratio between 6 and 9, as well as 8 and 12. (These are all the numbers in Nicomachus' "greatest and most perfect harmony.")

John Dee and the first use of the colon to compare equivalent ratios

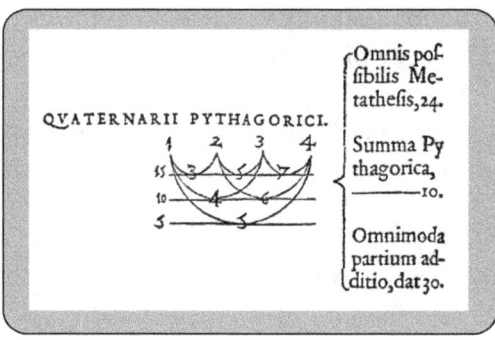

In his 1564 book, *Monas Hieroglyphica (Sacred Symbol of Oneness)* Dee explains the members of the "Pythagorean Quaternary" (1, 2, 3, and 4), multiply to 24, sum to 10 and the sum of all the pairs of combinations is 30.

In another chart, Dee cryptically shows the ratios 1:2, 2:3, and 3:4.

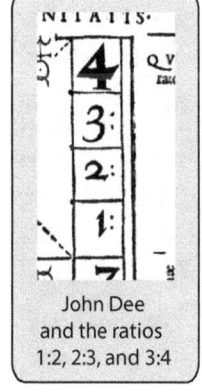

John Dee
and the ratios
1:2, 2:3, and 3:4

WILLIAM
OUGHTRED

In 1631, the English mathematician William Oughtred adopted Dee's use of the colon. However, Oughtred used the colon for the ratios and he used the double colon to compare ratios. (Oughtred also invented the slide rule and introduced the idea of using "x" as a symbol for multiplication.)

These ratios 1:2, 2:3, and 3:4 don't seem very related, but they are. These six equations demonstrate their amazing interconnectedness.

Whoa! Jump back jack. All these numbers are making me dizzy!

O.K. Let's transform these equations into something fun, something tangible: the 4 colorful puzzles that came packaged with this book.

$$\frac{1}{2} \times \frac{3}{2} = \frac{3}{4} \qquad \frac{1}{2} \times \frac{4}{3} = \frac{2}{3}$$

$$\frac{2}{3} \times \frac{2}{1} = \frac{4}{3} \qquad \frac{2}{3} \times \frac{3}{4} = \frac{1}{2}$$

$$\frac{3}{4} \times \frac{2}{1} = \frac{3}{2} \qquad \frac{3}{4} \times \frac{2}{3} = \frac{1}{2}$$

### *The 4 puzzle boards that accompany this book*

For those of you who don't like instructions, simply go ahead and try to figure out how all the 12 puzzle pieces fit on the four puzzle boards.

(Hint: Four pieces fit on each board, and when you're finished, all the printed numbers should be visible, so you can study the result.)

### *Example: Solving one of the puzzleboards*

Puzzle A

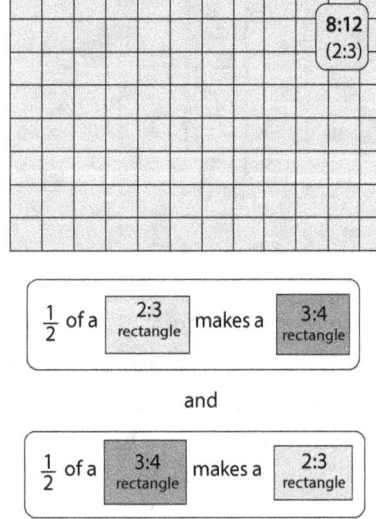

8:12
(2:3)

Find the red 8:12 rectangle labeled "Puzzle A."
Just underneath, it reads
"1/2 of a 2:3 rectangle makes a 3:4 rectangle."

$\frac{1}{2}$ of a | 2:3 rectangle | makes a | 3:4 rectangle

and

$\frac{1}{2}$ of a | 3:4 rectangle | makes a | 2:3 rectangle

Well, half of 8:12 rectangle would be a 6:8 rectangle. (There are actually two blue 6:8 rectangles provided. Use the one that is oriented vertically.)

Place this blue 6:8 rectangle on top of the left side of the red 8:12 rectangle, so all the numbers will remain visible.

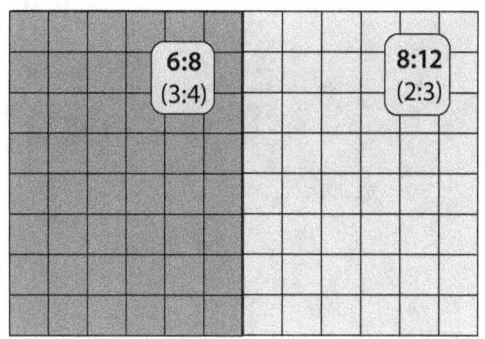

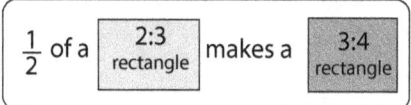

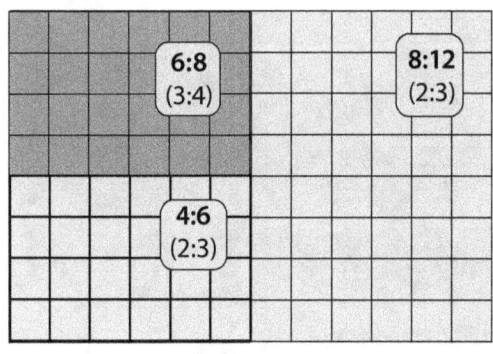

Also beneath the puzzle board it reads, "1/2 of a 3:4 rectangle is a 2:3 rectangle."

Well, half of that 6:8 rectangle would be the red 4:6 rectangle.

And half of that would be the little red 3:4 rectangle. That's it for this particular puzzle.

Notice that, by dividing in half, the rectangles switch from blue 3:4, to red 2:3, to blue 3:4, to red 2:3. This alternating pattern will continue, inwardly and outwardly, infinitely…

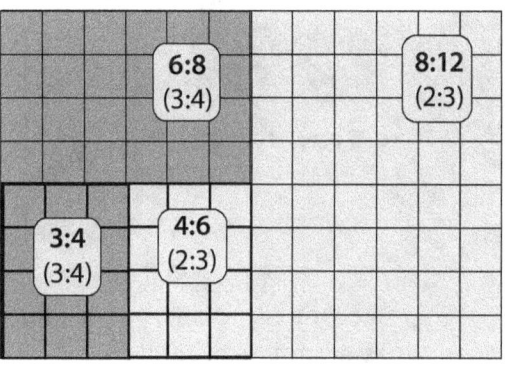

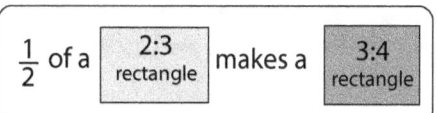

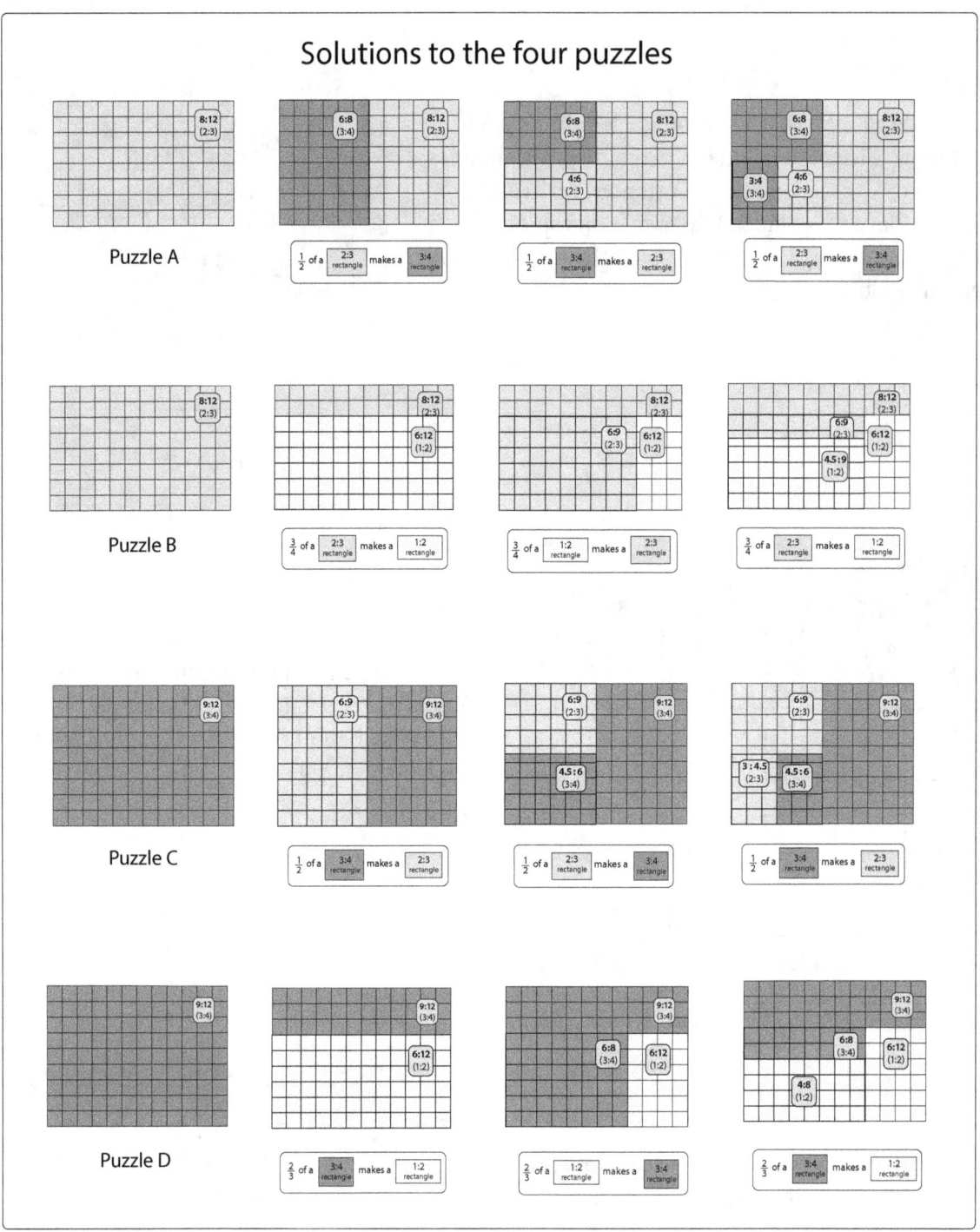

## Solutions to the four puzzles

Puzzle A

Puzzle B

Puzzle C

Puzzle D

Once you figure out the other three puzzles and contemplate your accomplishments, you will have a real feel for the close interrelationships among 1:2, 2:3, and 3:4.

You'll also notice that **most of these puzzle parts include the numbers 6, 8, 9, and 12** from Nicomachus' "greatest and most perfect harmony."

These puzzles help our eyes, hands, and mind understand Pythagoras' favorite ratios. Let's explore how we can actually *hear* these ratios!

PYTHAGORAS

## *Pythagoras and the symphony in the blacksmith shop*

Tradition holds that one day Pythagoras was walking by a blacksmith shop and was fascinated by the musical "clinking" sounds of hammers beating on anvils.

Entering, he observed:
big hammers made low-pitched sounds,
medium-sized hammers made somewhat higher-pitched sounds,
and small hammers made high-pitched sounds.

Thinking there might be a mathematical relationship between the sizes of the hammers, he went home and set up some experiments. He tested bells of different sizes, glasses filled with various amounts of water, strings of different tensions, and flutes of various lengths.

While it has since been proven that some of the experiments pictured in this 1492 woodcut (from a book by the Italian musician Franchinus Gaffurius) would not actually work to demonstrate what Pythagoras eventually discovered, the general principles are similar.

Instead of testing strings having various weights or tensions, is more likely that Pythagoras used a monochord, a stringed instrument like a lyre, but with only one string.

(IUBAL or JUBAL was called the Father of Musicians in the Bible. Philolaus was a follower of Pythagorean Wisdom who lived around 400 BC.)

## *Building a "Trichord"*

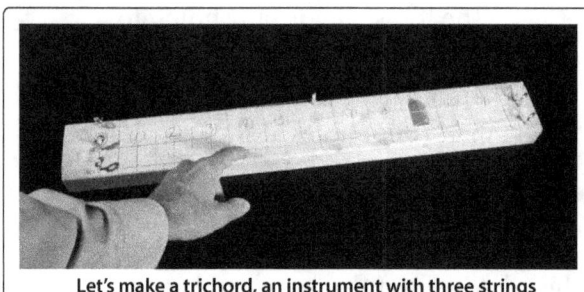

Let's make a trichord, an instrument with three strings

Reading about discord and harmony pales compared to actually hearing it, so let's make a simple stringed instrument.

The best way to judge harmonies is to hear them in a sequence. Instead of just a monochord, let's make a "Trichord," (with three strings) so we can hear a sequence of low, medium, and high notes.

For this you will need a wooden "2 by 4" which is at least 28 inches long.
(Actually any piece of wood 28 inches long and at least one-half inch thick will do.)

## Step 1: Marking and numbering the 12 notes

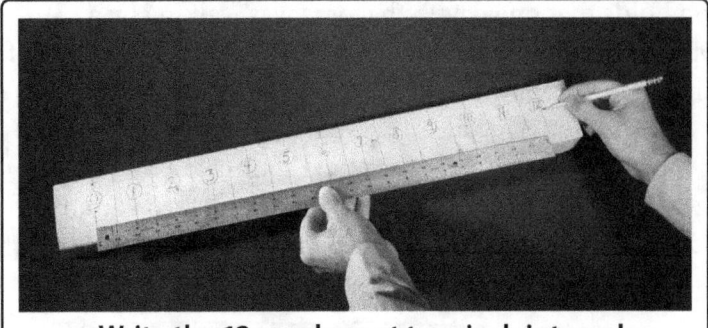

**Write the 12 numbers at two-inch intervals**

We'll be attaching 24-inch long strings to several of the screw eyes, so we have to leave ample room (two inches) at each end of the wood. First, draw a vertical "starting line" about 2 inches from the left end of the "2 by 4."

Starting there, lightly put a pencil line at each of the 24 inch-marks. Repeat this procedure a little higher up so you'll have two reference marks with which to make vertical lines. Next, draw in 24 vertical lines.

Now, at every "other" mark, write the numbers 1, 2, 3, … up to 12. You can erase part of the pencil line so the number is right on the line. (In other words, we have divided the 24 inches into 12 sections, each 2 inches wide.)

## Step 2: Making pilot holes for the screw eyes

To make the screw eyes easier to screw in, first we'll make pilot holes, using a hammer and a medium-size nail. (The nail should be about the thickness as the screw of the screw eye.)

Hammer the nail about 3/8 of an inch into the wood. Then use the hammer's claw to pull it out.

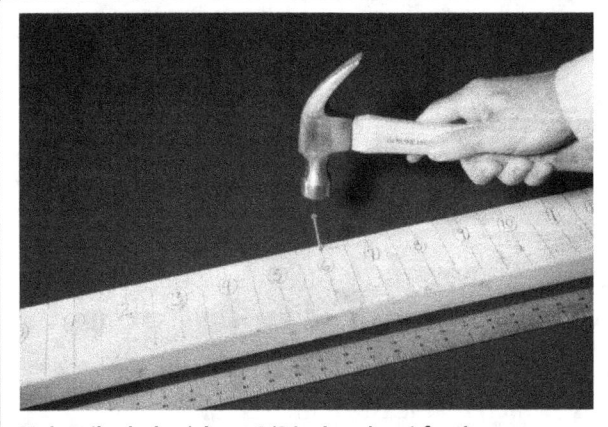

**Make pilot holes (about 3/8 inches deep) for the screw eyes**

The top string is only going to be "6 units" long (or 12 inches).

In other words, for the top string, make pilot holes at the "zero" line and at the "6" line.

For the lower two strings, put the pilot holes at the "zero" and at the "12" line (which is actually the 24-inch line, at the far right).

## Step 3: Orienting the screw eyes

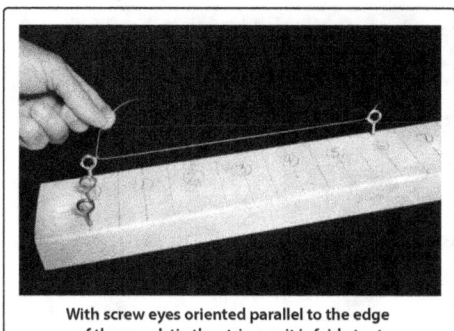

With screw eyes oriented parallel to the edge of the wood, tie the string so it is fairly taut

Now put screw eyes in each of the six pilot holes. *But don't screw them in all the way.* Only give them a few turns until they feel stable enough to stay upright in the hole.

Orient the screw eyes so they are all horizontal when viewed from above (In other words, parallel with the edges of the "2 by 4.")

## Step 4: Attaching the strings

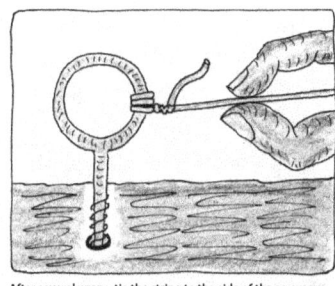

After several wraps, tie the string to the side of the screw eye

Next, we'll attach the strings (actually 50 pound test fishing line). Feed one string through the eyelet three times. Then make three or four overhand knots so the attachment is secure.

Feed the other end of the string through the "opposing" eyelet three times, and tie the end securely. Make this string as taut as possible, but without causing the screw eyes to lean too much in their pilot holes. (Snip off any extra string with scissors.)

Now, with a fingernail, slide the knot down to the bottom of the eyelet and give the screw eye one or two *clockwise* turns.

The string should start to wrap around the post under the eyelet. *Don't tighten too many turns just yet; only enough to get it started wrapping around the post.*

Perform the same "wrapping" technique on the opposing screw eye.

Push the knot to the bottom of the screw eye

Rotate clockwise so string wraps around the post of the screw eye

When plucked, the string should make a low "twang" sound. To give more tension to the string, we will twist the two screw eyes clockwise **at the same time** so they go lower into the pilot holes at the same rate. *But don't do that yet!*

Before we "fine tune" things, we need to tie on the bottom two strings to their respective screw eyes.

When we're finished, the top string will be a "6" sound.

The bottom string will be a "12" sound.

We will be varying the length of the middle string with a movable fret to perform various sound tests.

## Step 5: Tightening the bottom string

Using leverage to turn the screw eye clockwise, tightening the string

For our "final tuning," we'll start with the bottom string. Tighten its two screw eyes clockwise simultaneously until they start to hurt your thumbs (ouch!). For more leverage slip a nail (or screwdriver or pencil) through the eyelet and screw them in a bit more.

Don't stretch the string so much that it breaks, but enough that it makes a nice vibrating sound when plucked.

By now, all of the threads of the two screw eyes should be just about hidden in the wood. (Don't worry if the screw eyes start leaning towards each other a little bit, they will still be stable enough.)

## Step 6: Fine tuning the middle string

Now, here's where your musical ear comes in. Tighten the middle string so that it sounds identical to the bottom string.

This might take some fussing around, some loosening and tightening until you get it right, but with a little patience you can get the two sounds match up quite well. Again, it doesn't matter exactly what that sound is, as long as the two sounds match.

## Step 7: Fine tuning the top string

Finally, we'll tune the top string. As it's only six units long, put the triangular "movable fret" (provided) on the number 6 **of the middle string.**

Pluck the section of the middle string that is to the left of the "6." Tighten the top string until it matches that sound. (Remember, don't adjust the middle string, as it has already been tuned.)

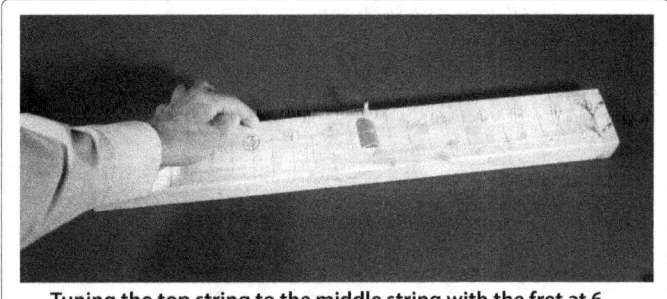

Tuning the top string to the middle string with the fret at 6

Your trichord is complete!

First, we'll use the "movable fret" to test various "pairs" of sounds. Then we'll move on to test "three-note sequences."

Incidentally, you can finger pluck, or use a guitar pick, or even carefully cut a guitar pick from the side of a plastic soda bottle.

### 6 and 12 make a harmonious pair

Position the fret **on the 5 of the top string.** Pluck the top string, then immediately afterward, the "full 12" bottom string. Not a very harmonious sounding duo, are they?

Now remove the fret from the top string so it's a "full 6." Pluck it, then pluck a "full 12" on the bottom string. Can you hear that sweet harmony? They sound great together!

This makes sense, because 6 is half of 12. The vibrating strings have similar "waves" running through them.

(In traditional music parlance, these sounds are said to be "an octave" apart. I'll explain where this idea of "eightness" comes from in a moment.)

Using this "6 and 12 harmony" as a reference sound, next we'll test various "three-note sequences" by incorporating the middle string.

### Three-note sequence tests

Pluck 6 (top string), then 7 (middle string), then 12 (bottom string). *Yuck.* That 7 doesn't harmonize very well with 6 and 12. Something sounds awry or "sour."

Pluck (6, 8, 12). Now this should sound like a really sweet trio. Pluck this sequence a few times and you'll hear 8 has a nice "open", "clear" sound that fits nicely between 6 and 12.

Pluck (6, 9, 12). That should sound great as well, but in a slightly different way.

Pluck (6, 8, 12), then (6, 9, 12), a few times and hear the differences. They are different but they both sound like nice, harmonious sequences.

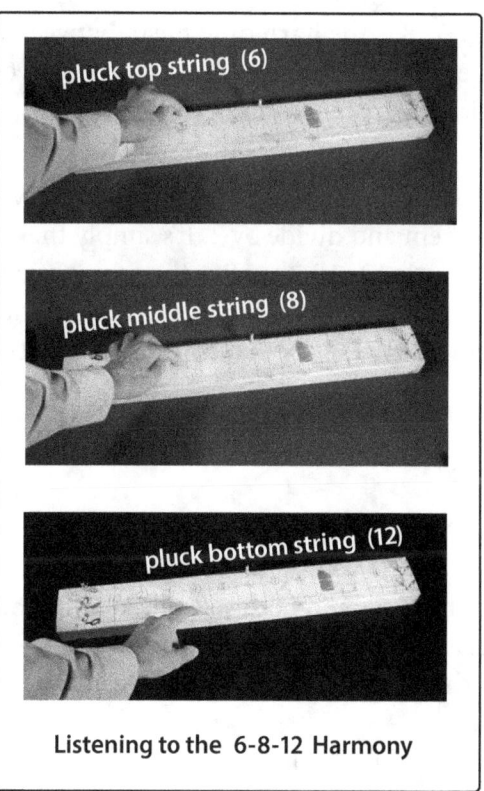

pluck top string (6)

pluck middle string (8)

pluck bottom string (12)

**Listening to the 6-8-12 Harmony**

Pluck )6, 10, 12). This sounds *yucky.* The 10 seems "sour" among "sweet" 6 and 12.

Pluck (6, 11, 12). An even *yuckier* sequence. The 11 is so far from 6 and so close to 12, the sequence just sounds awkward and unbalanced.

## Here's a revealing test:  6, 8 1/2, and 12

As the (6, 8, 12) and the (6, 9, 12) sound great, it seems logical that the (6, 8 1/2, 12) might sound just as nice. Well, give it a try.

Plucking 6, 8 1/2, and 12 is *yucky*. Compared to the sweet (6, 8, 12) and (6, 9, 12), this (6, 8 1/2, 12) sounds really sour or discordant (at least to most Western ears).

The reason the whole range from 8 to 8 1/2 to 9 does not produce concordance is that it's *precisely* the 8 and *precisely* the 9 which are important.

Even the ancients knew 8 is the **harmonic mean** between 6 and 12.

And they also knew 9 is the **arithmetic mean** between 6 and 12.

## What do these means mean?

To find the **harmonic mean** between two numbers, divide 2 by the sum of the reciprocals of the numbers.

1/6 and 1/12  sum to 1/4. When 2 is divided by 1/4, the result the 8.

8 is the **harmonic mean** between 6 and 12. Here's an easier way to see it: the distance between 8 and 12 is twice the distance it is between 6 and 8.

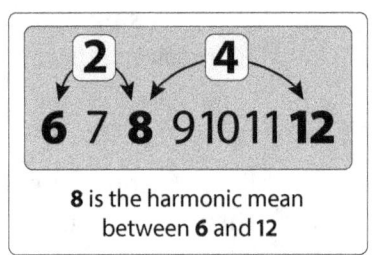

**8** is the harmonic mean
between **6** and **12**

To find the **arithmetic mean** between two numbers, add them and divide by 2. It's simply their average.

6+12=18, thus 9 is the arithmetic mean between 6 and 12. In other words, 9 is the same distance from 6 as it is from 12. That's why that trio sounds so stylin'.

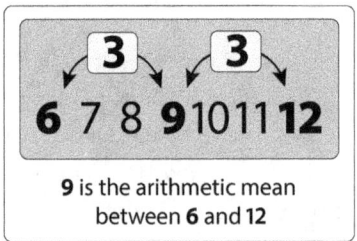

**9** is the arithmetic mean
between **6** and **12**

## Geometry = Music

The central idea is that with our Trichord we can both **see and hear** the most harmonious three-note sequences.
In this sense, Geometry and Music are the same.

How the numbers on our Trichord relate to the keys on a piano is a more complex story. So, novice musicians can skip over this next short chapter, but those already familiar with musical notation will find it interesting.

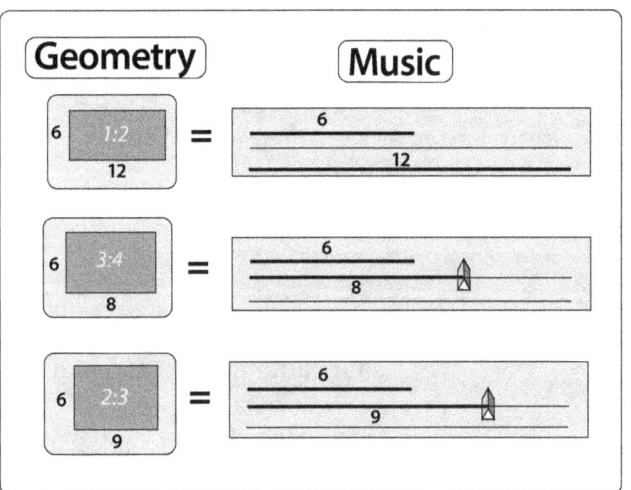

## *Equating our Tricord with a piano keyboard*

Why did we build the Trichord when these tests can be done quite simply on a piano? Answer: So we could see the "string length ratios."

These same "string length ratios" are inside the piano, but open the top of a piano, and you'll see they're challenging to locate.

Our trichord tests can be performed on the piano by using, for example,
6 = high C,   8=F,   9 = G,   and 12 = low C.

Plucking (6 and 12) on the Trichord would be similar to playing "high C" and "low C" on the piano.

"High C" is the eighth white key to the right of "low C," so the interval is called an "octave."

Plucking (6, 8, and 12) would be similar to playing "high C," F, and "low C" on a piano.

F is the fourth white key to the right of "low C," so the interval is called a "fourth." (F is the harmonic mean between the two C's)

Plucking (6, 9, and 12) would be similar to playing "high C," G, and "low C" on a piano.

G is the fifth white key to the right of "low C," so the interval is called a "fifth." (F is the harmonic mean between the two C's)

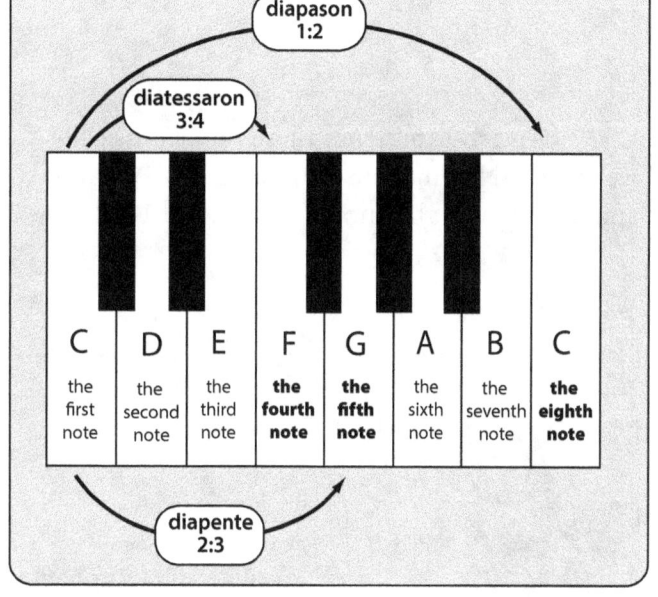

## *A caveat about judging musical sounds:*

There is really no such thing as a "bad sound" or "good sound." Some notes sound a little awkward or discordant when compared to other notes in the sequence or in a song. Other notes sound really "harmonious."

The judgment of "good" or "bad" is in the ear of the beholder. Western ears often appreciate different sounds than Eastern ears.

And even "discordant" notes are useful in songs. Indeed, in some musical passages, discord is composer's intent. Without discord, concord often has less impact.

*Conclusion*

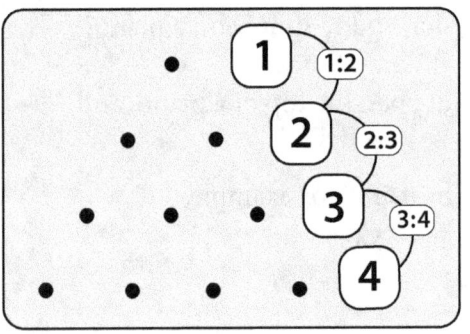

**Pythagoras** had this all this music, number, and means stuff figured out 2500 years ago.

He probably went through similar musical tests and came to the conclusion "music is mathematical."

Just as the ratios 1: 2, 2:3, and 3:4 are the essence of the tetractys, they are also the essence of harmonious sounds.

And to experience these ratios musically we can use **Nicomachus'** and **Boethius'** "greatest and most perfect harmony," 6, 8, 9, and 12.

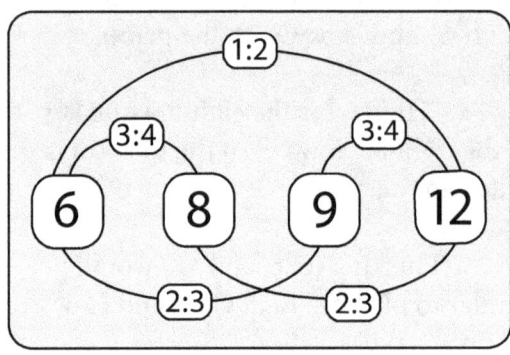

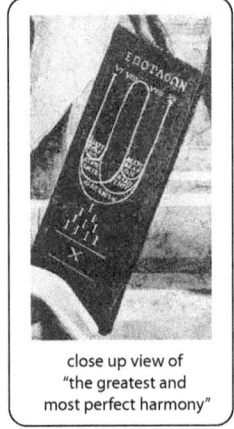

close up view of "the greatest and most perfect harmony"

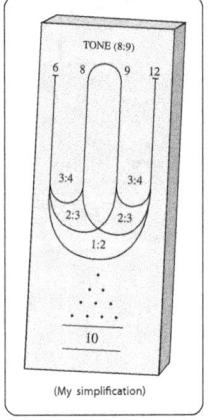

(My simplification)

This whole idea is what inspired **Raphael** to include 6, 8, 9, and 12 in his *School of Athens* fresco.

It's what excited **Leon Battista Alberti** and **Andrea Palladio** to design architecture with harmonic proportions.

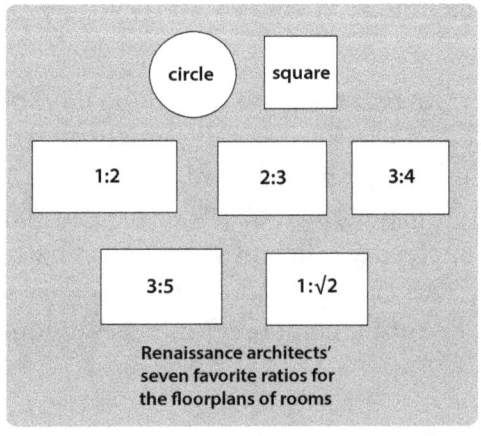

Renaissance architects' seven favorite ratios for the floorplans of rooms

It's what excited **Franchinus Gaffurius** in his early Renaissance music theories.

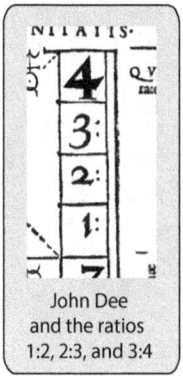

John Dee
and the ratios
1:2, 2:3, and 3:4

It's what fascinated the great mathematician John Dee in Elizabethan times.

We have explored how **number, shape, and sound** are interrelated.
In other words, we have seen how **Arithmetic, Geometry, and Music** are interrelated.

$$\frac{1}{2} \times \frac{3}{2} = \frac{3}{4} \qquad \frac{1}{2} \times \frac{4}{3} = \frac{2}{3}$$

$$\frac{2}{3} \times \frac{2}{1} = \frac{4}{3} \qquad \frac{2}{3} \times \frac{3}{4} = \frac{1}{2}$$

$$\frac{3}{4} \times \frac{2}{1} = \frac{3}{2} \qquad \frac{3}{4} \times \frac{2}{3} = \frac{1}{2}$$

**Arithmetic**

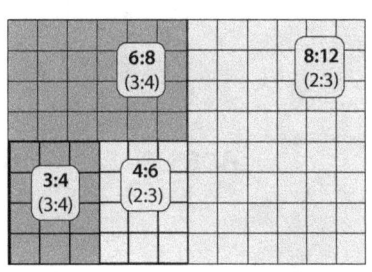

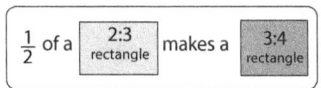

$\frac{1}{2}$ of a [2:3 rectangle] makes a [3:4 rectangle]

**Geometry**

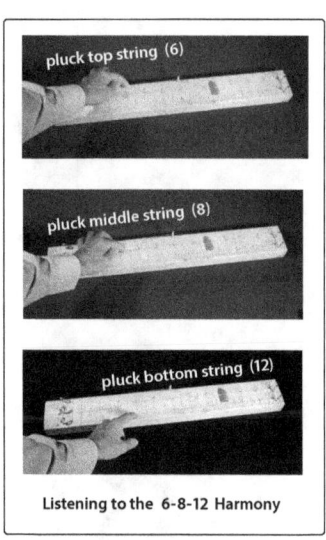

pluck top string (6)

pluck middle string (8)

pluck bottom string (12)

Listening to the 6-8-12 Harmony

**Music**

It all boils down to the tetractys invented by that genius Pythagoras over 2500 years ago. We thank you, *Pitagore*.

Now let's see if you, dear reader, can find some novel way of applying this "wisdom of the past" to new ways of making art, architecture, and music, today, and in the future.

# A Tetractys of Scholars

PYTHAGORAS

PLATO

VITRUVIUS

NICOMACHUS
(INTRODUCTION TO ARITHMETIC)

BOETHIUS

LEON BATTISTA
ALBERTI

ANDREA
PALLADIO

RAPHAEL

FRANCHINIUS
GAFFURIUS

JOHN
DEE

Here's a bonus puzzle for you.
Can you find three letter S's in a row (like SSS) somewhere in this book?
(Hint: One of them is a symbol)

Solution to bonus puzzle:

Theanswerispythagoras'ssecretcode,didyouguessit?

www.ingramcontent.com/pod-product-compliance
Lightning Source LLC
Chambersburg PA
CBHW081422170526
45166CB00010B/3436